The Social Media Starter Kit Workbook

Kerry Rego

Kerry Rego Consulting Books
Santa Rosa, CA

Kerry Rego Consulting Books
1011 2nd Street, Suite 100
Santa Rosa, CA 95404
www.kerryregoconsulting.com

Ordering Information:
Special discounts are available on quantity purchases by corporations, associations, and others. For details, contact the Kerry Rego Consulting Books at the address above or book@kerryregoconsulting.com

The Social Media Starter Kit Workbook / Kerry Rego. —1st ed.
ISBN 978-0-9906183-5-5

Table of Contents

Quick Tip Sheets and Templates
Visit http://bit.ly/SMSKtemplates for digital downloads

*These items are exclusive content to this workbook.

Social Media Strategy Template

Goals

Mission:

*SMART Social Media Goals (Specific, Measurable, Attainable, Relevant, Timeframe):

1.
2.
3.

Profit Centers/Value Propositions/Services:

1.
2.
3.

Target Audience (WHO)

WHO is your target audience? List them.
WHO are you (brand identity)? Know this answer.

Consider: age, stage of life, gender identity, socio-economic status, education, language, culture, needs, preferences, concerns, etc. Name them-Mom, Grandparents, Boss, Teens, Event planner, Purchaser, etc.

Target 1:
Target 2:
Target 3:

Messaging (WHAT)

WHAT do you want them to know?

WHAT do they want to hear from you?

Keywords:

Target 1:

Target 2:

Target 3:

Editorial Calendar & Content Planning (WHEN)

- Timely task list
- Social media weekly theme schedule
- Editorial calendar, monthly
- Editorial calendar, individual channel
- "15 Easy Content Topics"
- Keep in Master Social Media Document (digital download)

Channels to Use & Tactics (WHERE)

Technical details on each channel are later in Workbook.

Blogging Posts

- Topical tie-ins, what's in the news
- How to's
- Story/History of your organization
- Industry related events
- Vendors & partner features
- Testimonials
- Differentiation
- Address obstacles and challenges for clients, provide solutions
- Frequently asked questions

Email Content

- Upcoming events
- Testimonials
- How to, simple instructions
- Topical tie-ins, what's in the news
- Background bios, passion
- Videos

Facebook Posts

- Text (remember this needs to be conversational and easy to read)
  - Ask simple questions to engage audience
  - Hot topics
  - Contests for content
  - Little known facts, history

- Photos
  - Employees or volunteers at work
  - Behind the scenes
  - Product shots
  - Fun, human content, real people

- Links (worst performing, use with care, promote what is yours)
  - Valuable information found on the web
  - Program information
  - Conversion, calls-to-action
  - Promote other social channels

- Video posts (under two minutes)
  - Photos turned into a movie (Flipagram is a good app for this)
  - Saying hi from the team
  - Testimonials
  - Two options: upload video right to FB because your video will auto play in fans' feeds OR upload to Youtube then share as a link.

Twitter Posts (Tweets)

- Blogs
- Replies to others, conversation
- RT thought leaders in your industry
- Valuable links related to your industry
- Photos
- Videos
- Promote other social channels
- Shout-outs to other great accounts or followers

LinkedIn Posts

- Recycle blogs
- Topical tie-ins, what's in the news
- How to, simple instructions
- Studies, science, technical articles
- Personal posts as frequently as once per day relating to business, career, or your industry
- Professional brand posts 1 per week

Google+ Posts

- Post blogs
- Topical tie-ins, what's in the news
- How to, simple instructions
- Studies, science, technical articles

Instagram Posts

- Have guest Instagrammers for a day (extremely popular)
- Employees
- Services/products in action
- Behind the scenes
- 15 second videos
- Use of hashtags is imperative to gain a following. See hashtag guide http://bit.ly/hashguide

Pinterest Boards/Pins

- Create boards for each subject or use of services
- Specific to audience members
- Needs served
- Motivation
- Vendors/partners
- Other services that pair well
- Video
- Colors, seasons, holidays, cultural

YouTube Videos

- History/story of company
- User stories
- Testimonials
- Saying hi from the team
- Photos turned into a movie (Flipagram is a good app for this)
- POV (point of view) is powerful

Measure Performance

- Metric report
- Keep in Master Social Media Document (digital download)

Crisis Planning

- Reputation management tools
- Reputation management action list
- Crisis plan
- Keep Crisis plan in Master Social Media Document

Staff Responsibility

- Digital assets
- Keep Digital assets in Master Social Media Document

Reporting Procedure

- Daily, weekly, monthly, quarterly review
- Create easy to read reports of performance using Metric Report and provide to management at their preference in addition to hard data. Include in staff meetings.
- Verbal reporting should cover: goals, basic tactics, what behavior/metrics were observed, whether goals were met, what adjustments will be made in future.

Checklist

Have you put your social media channels:

- On marketing collateral?
- In your place of business?
- On your website?
- In your email signature(s)?

Social Media Categories

You will not need all of the tools listed here. I find that many people feel as if they are missing something in the social media lineup. I designed this graphic so that you can check off the ones you have and notice if you are heavy in one category. If that's the case for you, try more balance.

My favorite out of this bunch? Face-to-face. Your internet presence speaks for you when you're not there but nothing compares to being in the same room as another person.

Those with polka dots are traditional marketing (yes, Web is traditional) and plain are new media.

Social Media Tools By Category

This is not a complete list but a sampling of the most popular tools in each category. Some can be placed in multiple categories.

Blogs

Blogger
Wordpress
Twitter (microblog)
Tumblr

Email Marketing

Constant Contact
iContact
AWeber
MailChimp
MyEmma

Multimedia

YouTube
Vimeo
Flickr
Instagram
Google Hangouts
podcasts

Social Networks

Facebook
LinkedIn
Google+

Bookmark/Sharing

Pinterest
StumbleUpon
Reddit
Digg
We Heart It

Review

Yelp
TripAdvisor

Events

Eventbrite
Acteva

Location Based Services

Google Places/Maps/+Local
Get Listed
Foursquare/Swarm
SCVNGR

Gaming

Mafia Wars
Farmville

Blogs

Tools
Wordpress
Blogger
Tumblr
Typepad

Extras
Google Analytics – measuring & tracking
Feedburner – tracks subscribers
dlvr.it – auto publisher to SM
email newsletter signup box
Akismet - spam filter

Best Practices
- Comment on other industry blogs
- Promote blog on other social channels, recycle content
- Lead back to point of conversion
- Include calls-to-action
- Encourage comments
- Include SM share buttons
- Have at least one image per blog, preferably at top, under 800 pixels
- Use ALT tags on for enhanced SEO
- Include contact info

Timing
Start with once per month, work up to once per week
Monday is best, Tuesday is worst

Types of Blogs
Interviews
Screencasts
Editorial
Reviews
Definition
List
Personal/company profile
Case study
Comparison
Parody
FAQ
Special reports
Cheat sheets
Checklists
Infographics
Projects
Survey
Contests
Comics/jokes
Graphical freebie
Presentations
Podcasts
Webinars
Talking head video

Email Marketing

Tools

Constant Contact

iContact

AWeber

MailChimp

MyEmma

Timing

- Send emails no more than once per week, best is about once a month
- Best time to send is 5-7amPST. Mondays are best, Tues/Fri afternoon are worst. Success of weekend posts will depend entirely on the audience.

Best Practices

- Must be mobile friendly/responsive
- Have easy-to-spot colorful points of conversion (such as a buttons): Learn More, Sign up for Mailing List, etc.
- Promote other social channels (links)
- Include social share bar so they can share with friends
- Lead back to point of conversion
- Include calls-to-action
- Ask questions and encourage comments
- At least one image per email
- Include contact info
- Most important place for converting & sales!
- Speak to Target Audience's pain and needs
- Test before sending, click on every link to make sure they work correctly
- Headlines need to be eye catching and promise value

Facebook

Newsfeed

The "home" of Facebook
Majority of users stay here
Like your own page
This is where users interact with your page and information

Facebook Page Functions

Timeline/Timeshift (scheduling)
About
Map
Photos/Cover Photo
Videos
Offers
Events
Milestones
Private Messaging
Insights
Likes (also called Favorites)
Multi Admin levels
Act as page on Facebook
Advertising
Custom apps (not very useful)

Types of Posts

Text
Photo (includes profile & cover photos)
Video
Link
Event
Milestone

Best Practices

- Frequency-once per day, try 3x/week to start
- Post 3+ hrs apart or they will cannibalize engagement
- Links perform the worst, use sparingly
- 5th grade reading level means info needs to be light
- Schedule by using FB timeshift instead of 3rd party tools
- Promotion is necessary now, pay 4 goal-oriented content
- Users don't like hashtags, use sparingly

Best Times to Post*

6-8am weekdays (before work)
Lunchtime
7-10pm weekdays (after work)
Fridays are best day to post
**HOTLY debated area. Stats are all over.*

Image Sizes (pixels)

Cover 851 x 351
Profile 160 x 160
Custom Apps 111 x 74

Advanced Techniques

http://bit.ly/FBadnewsfeed

Twitter

Twitter Functions

Follow Friends, Authors, Thought Leaders, Clients
Create lists of people
Search by name, interest, keyword, hashtag, category, save searches
Post media
Post links
Private message
Trending subjects

Timing

- Up to 11x per day, try for at least one per day (Hootsuite scheduling)
- 6am-6pm Pacific (can reach Europe all the way through to Hawaii within those times)

Hashtags

- Bookmarks used to find posts and people around a particular subject
- Read more: http://bit.ly/hashtagarticle

Analytics

https://ads.twitter.com
Klout
Bit.ly
TweetReach
TwitterCounter

Tools

Hootsuite
Tweetdeck
Refollow
FriendorFollow – to find out who you are following that isn't following back
Twellow
Paper.li

Best Practices

- 140 characters is length of tweet, use 120 so it can be retweeted
- Interact with others as much as possible, gets most engagement
- Follow people that are in your industry, smart, & thought leaders
- Use hashtags to be found, 1-3 at a time
- Follow hashtag chats and trending topics
- Post multimedia
- DON'T DM (direct message) others unless you are already having a conversation or need to share contact info

Twitter's Help Site

http://bit.ly/twtrhelp

LinkedIn

Complete Profile

A profile photo
Industry and location
Summary
3+ positions with descriptions
Education
Skills 5+
50+ connections

*Users with complete profiles are **40x more likely** to receive opportunities on LinkedIn.*

Network (who is your target)

- Connect with 1st degree
- Previous colleagues & classmates
- Browse 1st degree connections connection lists
- Use People You May Know
- Use Advanced Search
- Join groups to find target market
- Save searches & create alerts (3 searches on free accts)
- Watch Network Interactions (use email notifications)
- Contact Who's Viewed Profile
- Follow target companies
- Ask for Introduction

ROI

Monitor interactions in groups
Before & After comparison of significant activity
Website referral traffic
Analyze Leads

Groups Creation

- Promote on other channels
- Create engagement strategy (start conversations, post relevant news, react quickly)
- Use announcements (1/wk)

Group Participation (50 max)

- Meet the host
- Find out about members
- Network in the "digital room"
- Send messages to members
- Promote your content in Groups
- Use Promotions or Jobs Feature

Best Practices

- Personal posts as frequently as once per day relating to business, career, or your industry
- Using a scheduler such as Hootsuite or Buffer allows for maximized postings
- Post here the same way as you would for email or blogs
- Business minded, scientific, serious

Company Page

Description w/ keywords
Pick industry
Location
Post updates
Insights
Post minimum 1x per week

Google +

Ways to Post

Text

Photo

Video

Link

Event

Poll

Circles

Allows for organization of contacts

Privacy settings (I don't trust this)

Communities

Join industry specific

Share your content

Create your own

Hangouts

Video chatting and collaboration

Up to 10 people can use at a time

Live or recorded to YouTube

Create engagement strategy (start conversations, post relevant news, react quickly)

Use announcements (1/wk)

Polls

Best Practices

- Put +1 button on your website http://www.google.com/+/learnmore/+1/
- Connect your YouTube account to your profile
- Personal profiles often work better to promote brand, more human
- Use hashtags for discoverability
- Network in the "digital room"
- Send messages to members
- Promote your content in Groups
- Use Promotions or Jobs Feature
- Tag people
- Use formatting
  - Bold-*this* becomes **this**
  - Italic – _this_ becomes *this*
  - Strikethrough - -this- becomes ~~this~~

Company Page

Description w/ keywords

Pick industry

Location

Post updates

Tied to your map listing

Pinterest

Terminology

Pin – an image added to Pinterest

Board – a set of theme-based pins

Pinning – the act of visually sharing content

Repin – reposting somebody else's pin

Pinner – the person who does the sharing

Why does it matter?

#2 behind FB for driving traffic

Exposure of content

Leads, Conversion

Social sharing

Optimize your profile

Company name (use business account for analytics)

Logo

Use descriptions with keywords

Links

Verify website

Great board cover images (230 x 307 pixels)

How to Get Followers

Promote through other channels

Pinterest follow button

Create boards that are subject specific (SEO keywords)

Engage with other pinners' boards

Answer questions in person with board as a resource

Best Practices

- Engaging visual content such as pictures, infographics, screenshots, avoid stock images
- Drive back to locations of conversion
- Pin content other than your own
- Use "The Pin Button" to speed up pinning http://about.pinterest.com/goodies/
- Add the "Pin It" button to your website so visitors can share your content
- Provide good descriptions in pins for better SEO
- Create a video gallery
- Use hashtags, keywords
- Host a contest
- Feature offline events
- Measure impact, Google Analytics, http://pinterest.com/source/yourdomain.com

Instagram

Terminology

Collage – using a 3rd party tool, collages are popular ways to show multi-images
IG or Insta – Instagram
DM – direct message, private comm
Regram – reposting somebody else's image
Selfie – image taken by someone of themselves
Tag – Putting @ in front of another account name, increases your post's reach and can forge a relationship with other users
TBT – Throwback Thursday, theme posting of images from the past

Why does it matter?

Traffic
60x the engagement of FB
Social sharing
59% of the worlds top brands use IG

Optimize your profile

Company name
Logo
Links

How to Get Followers

Promote through other channels
You MUST use hashtags http://bit.ly/hashguide
Instagram follow button
Interact with others, follow, follow back, and like
Answer questions in person with board as a resource
Use Explore tab to find new people

Best Practices

- Direct followers to links "in bio" due to no other clickable links in IG
- Balance fun and business images
- Connect to Facebook Page
- Use relevant, popular hashtags (not too many)
- Create 15 second videos
- Embed Instagram video on blog or website
- Don't post too much, laid back on IG
- Use a variety of filters, Normal is most popular
- Apps to get the most out of IG http://bit.ly/1gP4j7k
- Use FB Page to encourage those to enter photo contest (use hashtags)
- Reward followers
- Network, use Statigram
- Measure impact, Google Analytics

Additional Tools

Management/Dashboards

Hootsuite

Buffer

Composer.io

Tweetdeck

Dlvr.it

Monitoring/Analytics Tools

Google Alerts

Google Analytics

Klout

Social Mention

Crazy Egg

URL shortener

Bitly.com

Tinyurl.com

QR Codes

QRStuff.com

QRCode-Monkey.com

Image Sizes

http://bit.ly/SMsizes2014

Hashtags

http://bit.ly/hashtagarticle

Slang

http://bit.ly/KRCinternetslang

Digital Assets

	Username	Email Address	Password	Admins	Status (in use, closed, etc.)
Email Marketing					
Facebook					
LinkedIn					
Twitter					
YouTube					
Google+					
Pinterest					

Social Media House

Use this social media house to:

- establish your profit centers or value proposition (foundation)
- place the tools you are using in the various rooms
- determine what the tone, delivery style, and content belongs in each room of your social house
- Read a blog I wrote on this here: http://bit.ly/KRCsocialhouse

~ Time Out ~

Thanks for purchasing this companion workbook to The Social Media Starter Kit!

This is a supplement to the book and you will find more explanation of strategy within that volume.
Please feel free to download any template you need, the link is in the Table of Contents.

Let me know if there's any way I can help you because that's what I do.

My best,

Kerry Rego

@kregobiz

707-520-4572

KerryRegoConsulting.com

Trainer, Teacher, Speaker, Author, Columnist, and Unicorn Lover.

Helping People and Technology Together Change the World.

Timely Task List

This is an example of what your Daily, Weekly, Monthly scheduling of tasks will look like.

Daily

check messages (< 5 min x 2)

Choose one channel per day, examples (30 min)
Facebook-schedule, talk, watch competition
Pinterest-post, comment, look for inspiration
Blogs-read and comment on others
Youtube-view, research, post
LinkedIn-read, comment, check profile

Monthly

Metric report (20 min)
Youtube video record & edit (undetermined)
Email blast-write, edit, test, send (approx 1.5 hr)
Blogging (approx. 1 hr)

Example Goals *Specific, Measurable, Attainable, Relevant, Timeframe*

1 Increase website/shopping cart visits
2 Increase audience on social media channels
3 Improve brand awareness
4 Increase size of email list

Social Media Weekly Theme Schedule

Day	Theme	Content	Channel Focus
Monday	Back to School Monday	education classes	Facebook LinkedIn
Tuesday	Tool Tuesday	new tools	Twitter
Wednesday	Strategy	consult book	Google+
Thursday	Throwback Thursday Trivia Thursday	history facts via questions	Youtube
Friday	Fan Friday	testimonials case studies	Instagram Pinterest
Saturday	Silly Saturday	jokes	Blog
Sunday	Safety Sunday	security online safety	Blog

Profit Centers

classes	consulting
speaking	book

This is the one I use. It gives you an idea of how you can use it.

Social Media Weekly Theme Schedule
Blank

Day	Theme	Content	Channel Focus
Monday			
Tuesday			
Wednesday			
Thursday			
Friday			
Saturday			
Sunday			

Profit Centers

Theme Pick a theme for each day for variety.
Content What content can you share that aligns with profit centers?
Channel Focus Spend .5hr each day focusing on a specific channel.
This allows you intimate knowledge, building relationships, spotting trends, and to learn new tricks.

Editorial Calendar Monthly

	Theme for Month	Web/Blog	Email	Facebook	LinkedIn	Youtube
January '15	History/Story of Company					
February '15	Primary Service					
March '15	Educational Conference					
April '15	Community Service					
May '15	Secondary Service					
June '15	Client Success Story/Testimonial					
July '15	Differentiate					
August '15	Primary Service					
September '15	Communication					
October '15	Employees					
November '15	Secondary Service					
December '15	Vendor Story/Referral					

* Change the channels up top to fit your needs

* You may want to add rows to have 4-5 weeks per month.

Example themes come from http://bit.ly/KRCeasyblog

Editorial Calendar, Individual Channel

	TOPIC/TITLE	CONTENT/DETAILS	KEYWORD(S)	TARGET PERSONA(S)	OFFER/CTA
MONDAY					
Author:	12 Ways to Stay Healthy	Explain importance of cooking and exercise and how to benchmark your efforts to guarantee success	healthy cooking, exercise benchmarks	Parents	Web page to visit
Due Date:					
Publish Date:					
TUESDAY					
Author:					
Due Date:					
Publish Date:					
WEDNESDAY					
Author:					
Due Date:					
Publish Date:					
THURSDAY					
Author:					
Due Date:					
Publish Date:					
FRIDAY					
Author:					
Due Date:					
Publish Date:					
SATURDAY					
Author:					
Due Date:					
Publish Date:					
SUNDAY					
Author:					
Due Date:					
Publish Date:					

15 Easy Content Topics by Kerry Rego (excerpt)

http://bit.ly/KRCeasyblog

1. **Seasons, weather, and holidays.** How is your business affected by the seasons? If you are a tax accountant, first quarter looks very different for you and your clients than the rest of the year. Clothing retailers adjust their offerings based on the season. What are your seasons? Also, each day, week, and month celebrate something. If today is National Peanut Day and you sell peanut butter, talk about it!

2. **Busy time and quiet time.** Your communications are quite different when things are slow as opposed to when they are busy. What product or service would you like to sell more of during the slow season?

3. **Industry related events.** Most industries have annual conventions and educational events. When you return after attending one of these functions, what have you learned that you can share with your customers?

4. **Education.** Every industry has its changes and you are an expert in your field. What changes in your industry do your customers need to be educated about?

5. **Employee features.** You probably don't do it alone. Highlight the great team you have that helps you provide great products and service. Your team will get a boost in their morale due to recognition and your customers will learn more about the faces and families behind the product they are getting from you. It becomes personal.

6. **Vendors and partners.** The vendors you work with are great for a reason, tell us about it! Talk about why you choose to work with them. Those partners will be grateful for the free press and it will solidify your relationship.

7. **Case study and/or client success story.** Seeing how you've helped others will help your reader identify and apply the story of success to themselves.

8. **Testimonials and interviews.** Get client testimonials (especially when they are really happy!) and let their words to the talking. Video testimonials are the best. Talk to industry experts, your best customer, vendors, thought leaders. It takes content out of its normal context and provides a new way to talk about your subject matter.

9. **Product release.** What new product are you proud to be releasing? Give us some excitement by building it up. Let us know why you created it, what the demand was, how we can get it, and when it'll be ready.

10. **Hot topics.** If there is something exciting and dramatic going on with your industry or if it's in the news right now, weigh in with your opinion or break it down for the audience if the subject matter is confusing.

11. **History and story of your company.** Why did you start your business? What is important to you? People want to like the people they do business with and want to do business with people they like. Give them something to go on. They will tell your story for you when they recommend you to their friends.

12. **Differentiate yourself from your competition.** What makes you different? What is the benefit that your customer will get using your services? You can take this opportunity to clarify if there is any confusion about who does what. This is also very important to be able to express in all marketing scenarios.

13. **Identify obstacles and solutions.** These are the ones that you know like the backs of your hands. The problems they will encounter and the solutions that will help them overcome. Walk yourself through the typical client conversation, what problems do they experience?

14. **Survey.** Take a survey of your clients or prospects and release the information as a dataset. Tell them if you were surprised what you learned, what the community thinks, or what was reiterated.

15. **Frequently Asked Questions.** These are the 10 or so questions that everyone asks you. You answer them all the time. Since people continue to ask, you should continue to answer. These never get old. The trick is to express the question and answer in new ways.

Metric (Engagement) Report

	June	July	August	September	October
Website					
Google Analytics					
Visits					
Unique Visits					
bounce rate					
sources of traffic					
avg time on site					
#1 source of traffic					
Blog					
Subscribers					
Comments					
Facebook					
Fans					
Unsubscribes					
sources of traffic					
most viral post					
Twitter					
Followers					
RT's/Favorites					
YouTube					
Subscribers					
Channel Views					
Upload Views					
Interactions					
Google Places					
Impressions					
Actions					
Top search queries					

Reputation Management Tools

Proactive Tools

Website
Google Profile
Google Places
Facebook Page
LinkedIn Profile
Yelp
YouTube
Flickr
blog
Twitter
About.me
Slideshare.net
Online directories
Get Listed

SEO Tools

Google Analytics
Google Webmaster Tools
Google AdWords Keyword Tool

Monitoring Tools

Manual Vanity Search
Google Alerts
Yahoo Alerts
Social Mention
Brand Yourself
EdgeRank Checker
Klout
Tweet Angel
Technorati (for blogs)

Reputation Management Action List

- Perform a vanity search
- Own your own domain name
- Setup Google Alerts or other monitoring tools
- Setup Google Places/Google Business Page http://www.google.com/business as well as other business directories
- Respond to negative commenters found during vanity search
- Create a Crisis Plan
- Update Communications Policy
- Legacy Management/Process
- Assess Computer Systems security
- Review website for correct information, functionality, browser compatibility, update on a schedule to keep rankings high with Google

Blogs to show you how to do the action items: http://bit.ly/SMActionItems

Disclaimer: **You will not be able to accomplish all of this in one day.** I want to help manage your expectations. You will want to set aside time to accomplish these tasks on a regular basis (maybe once or twice a week) until you've worked your way through the list. Pace yourself. What you learn in this process about the online image of your name and your business will be worth all the effort.

Crisis Plan

Crisis Scenarios	Solutions
Website Hack	
Social Channels Hack	
Loss of Access	

Public Communications	Checklist
Statement for each crisis scenario	
Statement 1: Media	
Statement 2: Clients, Customers	

Emergency List of Contacts	List
Law Enforcement	
Fire	
First Responders	
City/County Officials	
Insurance Company	
Legal Advice	

Communicate with Clients	Checklist
In Person	
Mail	
Email	

Test Drills Performed, Dates	To Do List for Improvement

Marketing & Analytic Resources

Pew Research Center's Internet & American Life Project
Statistical information, studies, demographics, behavior
http://pewinternet.org

Hubspot
Case studies, signup by email to receive free white papers and ebooks on how to use social media and individual channels
http://www.hubspot.com/marketing-resources/

ComScore
Analytics on current web use
http://www.comscore.com/Insights/Blog
http://www.comscore.com/Insights/Presentations_and_Whitepapers

Mashable
New resource for social media and web, I am on their mailing list for a daily digest (it's a lot of information)
http://mashable.com/

Dan Zarrella, The Social Media Scientist
Statistics and data related to social media behavior.
http://danzarrella.com

Social Media Governance
The most complete listing of social media policies
http://socialmediagovernance.com/policies.php

Further Reading & Resources

Kerry Rego Consulting, category of blogs
http://bit.ly/krcrepmng

"What You Don't Know About Social Media CAN Hurt You: Take Control of Your Online Reputation" by Kerry Rego.
An in-depth look at reputation management and how to make sure you look good online with minimal effort. http://bit.ly/KRCRepBook

Pew Internet Research Reputation Management Report
http://bit.ly/lSnQM1

Removing Pages from Google, a Comprehensive Guide
http://selnd.com/lqZ2BS

Made in the USA
Las Vegas, NV
10 November 2026

57851004R00026